The Big Water Puddle:
A Look at the Water Cycle

**By:
Christen McCray-Moore**

The Big Water Puddle: A Look at the Water Cycle
Copyright © 2020 by Christen McCray-Moore

ALL RIGHTS RESERVED. Any unauthorized reprint or use of this material is prohibited.

No part of this book whether art or text may be reproduced or transmitted in form or by any means, electronic or mechanical, including photocopying, recording, or by any information storage and retrieval system without express written permission from the author except in the case of debrief quotations embodied in critical articles and reviews.
For more information, contact Armor of Hope Writing & Publishing Services, LLC, info@armorofhopewritingservices.com.

ISBN: 978-1-7336134-2-2

This publication is designed to provide accurate and authoritative information in regard to the subject make matter covered.

Illustrations: Afzal Khan
Editor: Denise M. Walker
Printed in the United States

One sunny, summer day, a little girl named Dee Dee and her family moved into a new home in a quiet neighborhood. The home was built next to a wildlife preserve. Dee Dee's home sat on an acre of land and all around it was lush trees, greenery, wildlife and trails that led down to a lake.

As soon as Dee Dee's family moved in, she had one thing on her mind. Dee Dee set out to go on an adventure! She was destined to see what was along the trails and which wildlife creatures lurked in the woods. Dee Dee wondered if there were ducks, alligators or turtles in the lake behind her house.

Once Dee Dee and her family got settled in, she asked her parents, "Can I go explore the trails and check out the lake?"

Her dad replied, "Yes!"

Her mom added, "But be careful, stay on the trail and watch out for the wild animals!"

Dee Dee was over the moon with excitement as she bolted out the door! She started down the trail. While skipping and hopping down the trail, she heard lots of birds chirping and saw blue jays flying through the air. Squirrels scurried along the path, jetting out to and fro. Then, all of a sudden, she stopped. She heard something rustle in the bushes! As she moved in closer to see what it was, Dee Dee felt and heard raindrops falling, slowly at first, but then they got heavier and faster.

Dee Dee thought, Oh no, I can't make it to the lake in all this rain. I better head home. In an instant, she sprinted towards home.

Upon her arrival, her mom said, "Thank goodness you're back. I was just starting to worry about you. Why don't you dry off? I was just about to start dinner."

Dee Dee replied, "Okay, Mom."

A few minutes later, Dee Dee's mom was in the kitchen getting ready to prepare dinner. She was making spaghetti and meatballs, Dee Dee's favorite! Her dad was in the living room, sitting in his favorite chair watching TV. Her baby brother was in his room playing with his toys.

Dee Dee found herself sitting and staring out the living room window, longing to go back outside but she couldn't because it was raining *cats and dogs*. DRIP, DROP, DRIP, DROP, DRIP, DROP! To Dee Dee, it had been raining for what seemed like forever. Big water puddles were beginning to form all over the yard.

Soon her mom came and stood beside her.

"You look sad, sweetheart. What's wrong?"

She pouted, "It is so BORING, Mom. Why is it still raining?" Dee Dee was upset that she couldn't go outside and continue exploring.

Her mom stated, "The rain will pass soon, and the big water puddles will dry up. Then, you will be able to go out and explore!"

Dee Dee nodded with sadness and answered, "Okay."

Moments later, her mom said, "You know I am a meteorologist, and I could explain the water cycle to you!"

"But how, what happens to the water on the ground? Where does it go? What is the water cycle?" Dee Dee asked with a curious expression.

"Well, there are tiny water droplets in our atmosphere. Some of them come from the stem and leaves of plants, known as transpiration. Some come from oceans and other bodies of water. When they are absorbed into the clouds, we say they have evaporated or changed from a liquid to a gas."

Dee Dee gave her mom a puzzled look and said, "Huh, I'm so confused."

Dee Dee's mom led her into the kitchen. She told her to look at the pot of boiling noodles on the stove. Her mom explained, "The heat speeds up the evaporation process.

The heat from the boiling noodles causes the water to become steam coming from the pot known as water vapor. When the steam comes in contact with the cooler air, droplets form, and we see a cloud.

The droplets evaporate quickly and change back to invisible vapor (gas)."

Suddenly, Dee Dee was in awe and surprised by her mom's explanation of evaporation. With excitement, she said, "Oh, wow. I thought the water just magically disappeared. Thank you for explaining it to me, Mommy."

"I was glad to explain it to you, sweetheart! Now, let me tell you all about the next phase of the water cycle, condensation."

For the next few minutes, Dee Dee's mom went on to explain, "When the water vapor/gas changes back to liquid water, condensation occurs. In this process, warm air rises to where temperatures are colder and some of the water vapor condenses into tiny water droplets. When the clouds cannot absorb anymore water, the clouds have to release the water and precipitation will fall as either, rain, snow, sleet or hail."

"Oh, just like it is doing now?" Dee Dee asked.

"Yes," her mom replied. "Come with me," Dee Dee's mom said, leading her back over to the window in the living room.

"The form of precipitation that we are experiencing now is rain, and it is coming from the big white, clouds and falling on our rooftops, cars, grass, trees, flowers and even us if were to go outside. I will tell you all about snow, sleet and hail at another time."

"Oh okay, that's cool. I never thought about how the rain helps us."

"Yes, rain has many benefits. Once the rain falls, some of it seeps back into the ground called infiltration, giving the ground the moisture it needs to help plants and animals grow."

"Oh, you mean it hides under the ground?"

"Yes, something like that," replied her mom as she led her over to the table and explained it in more detail.

Mom chuckled as Dee Dee looked up at her, "Sweetheart, the rest of the rainwater runs off, meaning it goes back into oceans, lakes, streams or other bodies of water. This starts the water cycle all over again! So you see the sun and the ocean are a huge part of the water cycle. If we didn't have oceans, the largest bodies of water, then the sun would not be able to dry up large amounts of water or aid in the process of evaporation very well, which in turn would hinder the processes of condensation and precipitation."

Dee Dee was fascinated. She asked, "Mom, did you learn all this working as a meteorologist?"

"Yes, Dee Dee! And guess what, it looks like the large water puddles have dried up!"

"You mean evaporated, right?" Dee Dee corrected.

"Yes, that is right! You are so smart!"

Just then, the doorbell rang. Dee Dee's dad opened it.

"Your friend is at the door, Dee Dee!" he yelled.

"Great! Can we go outside and explore now that we have a "break" in the water cycle?" Dee Dee turned to ask her mom.

"Sure," Mom stated.

Dee Dee walked to the door and began sharing with her friend what she had just learned. Her friend frowned, "Huh, what do you mean by that?"

Dee Dee said, "Come on, I'll explain it after we get outside."

As her friend sat on the porch listening, Dee Dee did her best to explain all that she had learned about the water cycle, "You see, when water on Earth's surface dries up, we say it has evaporated. The tiny water drops in the air are absorbed into the clouds called condensation. When the clouds get so full, they can't hold anymore water, it rains. Sometimes water goes into the ground, and other times it runs off into another body of water, like the ocean. Without the oceans, it would not be possible for the water cycle to work the way it does. So it is important to take care of our oceans."

"Oh wow, how did you learn all that?"

"My mom just taught me. So are you ready to go explore with me?" Dee Dee asked with a smile, as she dashed away with her friend.

Comprehension Questions:

- Who is the main character?
- What is the central or main idea?
- How does Dee Dee's mother help her to understand the water cycle?
- What is the conflict?
- Summarize the steps of the water cycle.

Christen McCray-Moore is a wife, mother, educator and author. Christen loves children. She has served as an educator at both the elementary and middle school level for over 20 years. Her favorite subjects are science and math. Christen's love for teaching children and science prompted her to start writing children's books. The goal of each book is twofold, to provide a story narrative as well as reinforce science standards that will enhance student achievement.

www.ingramcontent.com/pod-product-compliance
Lightning Source LLC
LaVergne TN
LVHW071027070426
835507LV00002B/58